Write It Yourself!

CE Marking Documentation:

Sell Your Devices in Europe

Checklists and Traceability Matrices for
- Electronics,
- Low Voltage Devices, and
- Machinery

A GUIDE FOR SMALL COMPANIES (AND BIG ONES) TO DOCUMENT THEIR DEVICES IN COMPLIANCE WITH THE EUROPEAN DIRECTIVES FOR CE MARKING.

Write It Yourself: CE Marking Documentation Checklists and Traceability Matrices for Electronics, Low Voltage Devices, and Small Capacity Machinery. Copyright © 2015 All rights reserved. No part of this book may be reproduced or transmitted in any form or by any means, electronic or mechanical, including photocopying, recording, or by any information storage and retrieval system, without written permission from the publisher..

Contents

What is CE Marking?..5

    The Benefits of CE Marking 7
    The Documentation Method for CE Marking..7
    The European Directives Referenced in This Manual: 9
    What is Covered; What is not Covered 11
    Documents Required for CE marking..15
    Documentation Overview: Risk Assessment..17
    Documentation Overview: Declaration of Conformity 19
    Documentation Overview: Device Manual..21
    Documentation Overview: Technical File/Technical Construction File..........27
    Declaration of Conformity Checklist Sample 37
    User Manual Requirements Checklist/Traceability Matrix...........................47
    The Technical File/Technical Construction File (TF/TCF) 71
    Machinery Technical Construction File Traceability Matrix 73
    Low Voltage/EMF Technical File Traceability Matrix.....................................74
    The Electrical Component File 77
    Terms 81

What is CE Marking?

The letters "CE" are the abbreviation of the French phrase "Conformité Européene," which in English means "European Conformity". "CE Marking" is now used in all EU official documents. In American lingo, "CE Branding" and "CE Mark" are terms used synonymously with "CE Marking," but the correct term is "CE Marking" and that is the term that should be used in any documentation where it is mentioned.

CE Marking on a product is a manufacturer's declaration that the product complies with the essential requirements of the relevant European health, safety and environmental protection legislation. This legislation assures the safety of European consumers and creates a standard of reliability for products manufactured in the European common market. The legislated requirements for CE Marking are put in place by the Product Directives. These Product Directives contains the essential requirements and/or performance levels to which the products must conform.

In reading Product Directives, you will see references to "Harmonized Standards." Harmonized Standards are the technical specifications (European Standards or Harmonization Documents) which are established by several European standards agencies (CEN, CENELEC, etc.). In other words, if your company wants to sell a product to Italy, Belgium, and England, instead of checking requirements and standards for each country separately, you can refer to the Harmonized Standards and safely use them for legally marketing your device to all three countries.

The purpose of this book is to offer assistance in the *documentation* necessary for CE Marking. Therefore, this introduction on CE Marking is merely cursory. If you are producing a prototype for sales to Europe, or if you want to open a new sales channel to Europe, your products must be tested and measured to assure that they comply with CE Marking requirements. That information falls outside the parameters of this book.

But if you are ready to produce the documentation required for CE Marking, this book can help you by offering you checklists and traceability matrices to help you meet all CE Marking requirements.

The Benefits of CE Marking

- CE Marking is an assurance from the manufacturer that a product may be legally placed on the market within the EFTA & European Union (EU) single market (total 28 countries).

- CE Marking ensures the free movement of a product within the EFTA/EU/and single market countries.

- CE Marking permits the withdrawal of the non-conforming products by customs and enforcement/vigilance authorities.

- Newer requirements are placing greater emphasis on CE Marking as a necessity to gain entry into EFTA/EU markets

The Documentation Method for CE Marking

This manual is designed to be used where the manufacturer of the device is seeking CE Marking acceptance in good faith. This manual is designed as a guide for the documentation side of CE marking acceptance. It is not designed to walk a company through the entire acceptance process.

Unless your device is incredibly simple, this manual recommends and provides guidance on five documents needed for acceptance, four of which are required by the Directives, and a fifth that forms part of good practices for companies producing electrical/electronic devices.

Use of this manual cannot guarantee CE Marking acceptance. The manufacturer must assure that the device itself is in compliance with the Directives. The documentation should then follow.

But if the manufacturer has put the time and energy into keeping an organized document profile of the device, this guide should assist in helping the writer organize the information that is required and put the appropriate information into the correct document.

The European Directives Referenced in This Manual:

- DIRECTIVE 2004/108/EC OF THE EUROPEAN PARLIAMENT AND OF THE COUNCIL of 15 December 2004 on the approximation of the laws of the Member States relating to electromagnetic compatibility, and repealing Directive 89/336/EEC referred to in this Manual as "The EMF Directive".

- DIRECTIVE 2006/42/EC OF THE EUROPEAN PARLIAMENT AND OF THE COUNCIL of 17 May 2006, on machinery, and amending Directive 95/16/EC (recast) – referred to in this Manual as "The Machinery Directive".

- DIRECTIVE 2006/95/EC OF THE EUROPEAN PARLIAMENT AND OF THE COUNCIL of 12 December 2006 on the harmonization of the laws of Member States relating to electrical equipment designed for use within certain voltage limits (codified version) – referred to in this Manual as "The Voltage Directive".

Therefore, for brevity's sake, in this Manual, the three Directives cited above will be called the EMF Directive (electronics), the Machinery Directive, and the Voltage Directive, respectively.

What is Covered; What is not Covered

The EMF Directive Includes and Excludes the Following:

The EMF Directive includes the following: any finished appliance or combination thereof made commercially available as a single functional unit, intended for the end user and liable to generate electromagnetic disturbance, or the performance of which is liable to be affected by such disturbance.

The EMF Directive excludes the following:
- Aeronautic equipment
- Defense equipment
- Licensed radio equipment (including amateur)

All of the above are all governed under different legislation.

The Machinery Directive Includes and Excludes the Following:

Machinery that is used outside of its intended use is not covered by either the Machinery Directive or this Manual.

Machinery included in the scope of the Machinery Directive:
- Machinery;
- Interchangeable equipment;
- Safety components;
- Lifting accessories;
- Chains, ropes and webbing;
- Removable mechanical transmission devices;

Machinery that is excluded (*Summary*):
- Weapons, including firearms, which are governed under different legislation,
- Certain types of agricultural and forestry tractors,
- Machines used to lift people that were not specifically designed to lift people,
- Partly completed machinery,
- Machines whose risks are increased by being used in certain environments or conditions, as stipulated by local governments,
- Machines determined to be high-risk by legislation of member communities

Further Machinery excluded from the scope of the Machinery Directive (*Detailed*):

- Safety components intended to be used as spare parts to
- replace identical components and supplied by the manufacturer
- of the original machinery;
- Specific equipment for use in fairgrounds and/or amusement parks;
- Machinery specially designed or put into service for nuclear purposes which, in the event of failure, may result in an emission of radioactivity;
- The following means of transport:
    - agricultural and forestry tractors for the risks covered by Directive 2003/37/EC, with the exclusion of machinery mounted on these vehicles,
    - motor vehicles and their trailers covered by Council Directive 70/156/EEC of 6 February 1970 on the approximation of the laws of the Member States relating to the type-approval of motor vehicles and their trailers (1), with the exclusion of machinery mounted on these vehicles,
    - vehicles covered by Directive 2002/24/EC of the European Parliament and of the Council of 18 March 2002 relating to the type-approval of two or three-wheel motor vehicles (2), with the exclusion of machinery mounted on these vehicles,
    - motor vehicles exclusively intended for competition,
    - and
    - means of transport by air, on water and on rail networks with the exclusion of machinery mounted on these means of transport;
- Seagoing vessels and mobile offshore units and machinery installed on board such vessels and/or units;
- Machinery specially designed and constructed for military or police purposes;
- Machinery specially designed and constructed for research purposes for temporary use in laboratories;
- Mine winding gear;
- Machinery intended to move performers during artistic performances;
- Electrical and electronic products falling within the following areas, insofar as they are covered by Council Directive 73/23/EEC of 19 February 1973 on the harmonization of the laws of Member States relating to electrical equipment designed for use within certain voltage limits ($_3$):
    - household appliances intended for domestic use,
    - audio and video equipment,
    - information technology equipment,
    - ordinary office machinery,
    - low-voltage switchgear and control gear,

- electric motors;
- The following types of high-voltage electrical equipment:
  - switch gear and control gear,
  - transformers.

Further exclusions from this Manual:
- Pneumatically driven machines
- Machines with combined electrical and pneumatic drives

The Voltage Directive Includes and Excludes the Following:

The Voltage Directive includes electrical equipment. "Electrical equipment" means any equipment designed for use with a voltage rating of between 50 and 1000 V for alternating current and between 75 and 1500 V for direct current.

The following items are excluded from the Voltage Directive:

- Electrical equipment for use in an explosive atmosphere
- Electrical equipment for radiology and medical purposes
- Electrical parts for goods and passenger lifts
- Electricity meters
- Plugs and socket outlets for domestic use
- Electric fence controllers
- Radio-electrical interference
- Specialized electrical equipment, for use on ships, aircraft or railways, which complies with the safety provisions drawn up by international bodies in which the Member States participate.

Documents Required for CE marking

The reason I combined checklists for EMF, Machinery, and Voltage Directives is that many devices come under the jurisdiction of all three Directives. But if you have a device that is covered by only one of the Directives, or two of the Directives, you simply cross out the non-pertinent columns. You also have permission of the publisher to scan or photocopy the check lists and modify them as needed for your project(s). You can scan them into a PDF, export them from a PDF engine into a word processor, and annotate them to suit your project. You cannot resell them, under the laws of copyright protection, or pass them out in any paid venue.

There are four documents that are specified by the European Directives. Written properly and assembled together, these documents can meet the minimum requirements as laid out in the three Directives:

- Risk Assessment
- Declaration of Conformity
- Manual
- Technical Construction File

In reality, the first three are usually part of the last one. For Machinery, the Technical Construction file encloses all the documents. But it is also a document in itself, as it is the collection of the technical drawings, in addition to the other documentation.

I would add a fifth document for electronics/electrical devices, including electrical machinery:

- The Electrical Component List

This document, as well, should be included in the Technical Construction File.

Documentation Overview: Risk Assessment

Chronologically, the Risk Assessment comes first in the documentation trail. The management of risk is part of Quality Analysis and Control. Risk assessment requires calculations of two components of risk: (R), the magnitude of the potential hazard, and (P), the probability that the hazard will be triggered.

No project is risk-free. The goal of risk assessment is to control risk and to assure that remedies are in place to mitigate risk or to neutralize hazards if they are triggered.

Risk is charted in an X/Y table format. The rows describe the probability factor: Frequent, Likely, Possible, Rare, Unlikely. The vertical column on the left describes the severity factors: Catastrophic, Severe, Moderate, Minor. The numerical rankings go from 1, most acceptable, to 3, unacceptable. A risk that is both likely and Catastrophic, for example, is a level 3 risk, unacceptable. A machine that explodes every Friday is not acceptable. Then again, a moving part that could pinch the fingers of the operator can be easily mitigated with a warning sticker and a first aid box nearby.

Risk Assessment/Risk Category						
Severity of harm || **Probability of Harm**					
		Frequent	Likely	Possible	Rare	Unlikely
Severity	Catastrophic	3	3	3	2	2
	Severe	3	3	2	2	1
	Moderate	3	2	1	1	1
	Minor	2	1	1	1	1

*Table taken from *IEC International Standard*, IEC 61010-1, Edition 3.0 2010-06

Ideally, the risks should all be level 1. Level 1 risks sometimes require a documentation mitigation strategy. That is, the documentation must explain how the risk is controlled: by lock out, by training, by warning labels, etc. If level 2 risks occur, then there *must* be a documented mitigation strategy.

It is not the writer's job to establish risk values, merely to document the results of the risk assessment. Nevertheless, typical risks include, but are not limited to, the following:
- Pinch points,
- Mechanical hazards
- Crushing hazards
- Chemical hazards
- Tipping hazards
- Shock hazards
- Hot Surfaces
- Lifting hazard
- Misuse hazards
- Tripping hazard
- Burn hazards

A template for Risk Assessment is provided later in this manual. Refer to the Table of Contents.

Documentation Overview: Declaration of Conformity

The Declaration of Conformity (DoC) represents the entity doing business with the European market. The DoC is a signed document providing assurance that the product meets the requirements of the directive(s) that apply to it. It identifies the responsible person who makes the guarantee that the specified device is qualified to bear the CE marking. The Declaration of Conformity must be traceable back to either the company that makes the device or the company that markets the device to Europe. It cannot be signed by a Notified Body or any third party service used in the production or qualification of the device.

The signer of the Declaration of Conformity must be the person with the authority and responsibility to legally and contractually affirm on behalf of the manufacturer or European distributor that the device meets all CE marking requirements; therefore the signer must be expert on the design and operation of the device and on the processes that have qualified the device. In one sense, the signer of the Declaration becomes responsible for the commitment made that the device meets all CE marking requirements. In another sense, the signer makes the device company or the marketing company responsible.

Each product directive requires a Declaration of Conformity, with each having its own set of requirements for what the Declaration must contain. The machinery directive is the most thorough, but all directives require the following:

- Name/address of manufacturer (and of responsible person where applicable)
- Model and/or serial number of equipment
- List of relevant directives
- Declaration statement
- Name and position of person signing
- Signature
- Date

If a Notified Body is required to be involved in the CE marking process, then their details will also need to appear on the Declaration.

The DoC for the Machinery Directive must be shipped with the device and it must be issued in the language of the end user (i.e. the same as the manual). For other directives the original language used must be one of the European Community languages. In addition, the machinery directive requires the name and address where the complete technical documentation for the product is stored, and this address must be within the European Community.

One device gets one Declaration of Conformity. If the device is governed by two directives (for example the machinery directive and the voltage directive), the requirements for both are combined into a single document. A checklist for the three combined directives appears later in this book. If your device does not fall under all three directives, you may edit the checklist or simply cross out any non-pertinent sections. You may scan it and edit it to tweak it for your device. But you may not sell copies or hand them out in any paid venue, per copyright law.

Documentation Overview: Device Manual

The manual must include safety, operation, and basic maintenance.

The priority of the manual is safety for the operators of the device. Operation and basic maintenance must be adequately covered. These tasks can later be expanded in a user's manual and technical manual for the machine for any commercial clients. But for the purposes of CE Marking, the manuals must include at least the bare bones of operation and maintenance, particularly startup information. But safety must be thoroughly covered.

The manual must include graphics of all safety stickers, labels, and engraved warnings, with explanations and definitions. To meet the Machine Directive, the device must have a nameplate affixed to it, with specifications included (See the checklist.) This nameplate should be documented in the manual: a graphic of the nameplate and a readable list of the included specs.

The following sample outline of a manual does not guarantee compliance with CE Marking, but it is consistent with the requirements for CE Marking. The sample outline is a generic tool designed to help writers organize information. Some headings do not apply to all devices and must be omitted. Some devices will require more categories than those listed below. Many headings will require the writer to add sub points, based on the complexity of the device or its unique requirements. Nevertheless, the following basic outline has been used successfully in creating manuals the meet requirements for CE Marking. Some headings will need to be replaced with specific task names or device features from the machine being documented. Adjust it as necessary.

Publication History

The publication history summarizes major changes to this document. Explain revision numbering as your company applied revision numbering. Include a table to show Revision number, a description of change to a document, and the date that change was released.

Terms and Conditions for Use

1.A Warranty. Limitation of Liability. "COMPANY NAME. warrants the Products will, upon delivery to Customer, conform to the description and specifications set forth herein and will be free from defects in material and workmanship for one year after installation."
1.B Define Warranty Limitations And Exclusions
1.C Define Patent, Trademark, And Copyright Limitations And Exclusions.
2.A Define Implied Patent And Copyright License. Exclude Any Changes Made To Product From License Agreement.
2.B Define Patent And Copyright Indemnification Rights And Restrictions
3. Define Security Interest. .
Define Any Other Limitations Upon Warranty, License, Copyright, And/or Safe Use Of The Device.

Chapter 1 - Machine Description and Specifications

1.1 Description
Overview of function of machine. Include graphic from photograph or line drawing.
1.1.1 Major Components
1.1.2 Interface and Software Overview
1.2 Operation Overview – Describe the operator's tasks.
1.2.1 "Best Path" Overview from Start to End. Provide a step-by-step narrative on the main objective of the device, assuming that it is being operated in perfect conditions.
1.2.2 Most Common Operational Tasks Overview – Provide a step-be-step narrative of the steps that the operator walks through to complete the most common tasks of operation, assuming ideal conditions.
1.3 Nameplate and Specifications
1.4 Additional Specifications
1.5 Layout and Dimensions (Express units in both US and European units of measure)

Chapter 2 - Safety Instructions

2.1 General Precautions – All information regarding adequate training, moving parts, electrical shock, burns, pinching, all other hazards, all required certifications, all lock-out measures.

2.2 Safety Symbols

2.3 Disclaimer of Liability

2.4 Warnings and Cautions – include all safety, warning, and caution symbols that appear on the device and in this manual in this section.

2.5 Safety Systems – describe any built-in or onboard safety systems on the device.

2.6 System Run Status – List alarms and other triggers that prevent operation

2.7 Danger Areas – List all areas of the device that present dangers to operators.

Chapter 3 – Delivery and Setup

3.1 Facility Requirements – List all environmental/facility requirements for safe operation of the device.

3.2 Operator Station – List all requirements for the operator station(s)

3.3 Delivery – List all requirements for receiving and accepting the device, including examination of bill of lading, invoices, certificates, and packing slips.

3.4 Unpacking Equipment – List all precautions, including the necessity of assuring that all parts have arrived.

3.5 Assembly and Setup – List all warnings and steps to take to set up the device. Include a blank space to record the device serial number(s). List all electrical/electronic connections that must be made: include illustrations.

3.6. Startup – Include a step-by=step procedural on powering up the device and starting it for operation.

Chapter 4 – Software

4.1 Software Overview – Describe the functions of the software.

4.2 Primary Screens and Their Functions – Show the screens. Identify all GUI functional items. Include instructions for the most significant or most commonly used procedures.

4.3 Diagnostics – Show the screens. Identify all GUI functional items. Include instructions for the most significant or most commonly used diagnostics.

4.4 File Menu Toolbar – List the purpose of each toolbar menu and the function of each menu item.

4.5 Function Menu – List the purpose of each screen menu and the function of each menu item.

4.6 Alarms and Faults – List each alarm that the device software will display and the actions it will require.

4.7 MORE – Any additional software information pertinent to the device.

Chapter 5 – Pneumatic Controls

5.1 Connecting to system air – include operating air pressure limits
5.2 Regulating air in the device
5.3 Manifold #1
5.4 Manifold #2

Chapter 6 – Electrical Controls

6.1 Control Panel #1 (Replace with specific name of panel, such as "Front Panel". Include photos or drawings.)
6.2 Control Panel #2 (Replace with specific name of panel, such as "Security Panel". Include photos or drawings.)
6.3 Onboard Controller #1
6.4 Onboard Controller #2

Chapter 7 – Automation Controls

6.1 PLC Inputs and Outputs
6.2 Sensor Types, Functions, and Locations

Chapter 8 – Operator Tasks

This Section includes all procedures for device operation. This section must include step-by-step procedures of the critical tasks necessary for the device to achieve its objectives. After acceptance, this section can be customized for specific clients, expanded with additional tasks that can be performed with the device, and even broken out into a separate user manual.

Chapter 9 - Cleaning Procedures

9.1 Cleaning Overview
9.2 Cleaning Procedures - General
9.3 Cleaning Procedures – Specific Areas or Priority Areas

Chapter 10 - Maintenance Guide

10.1 General Guidelines
10.1.1 Once per week – Maintenance tasks that are weekly.
10.1.2 Once per month – Maintenance tasks that are monthly.
10.1.3 Once per quarter – Maintenance tasks that are quarterly.
10.1.4 Once per year – Maintenance tasks that are annual.
10.1.5 Other – Any other periodic maintenance.
10.2 Shipping/Transporting the Machine – Supply directions for the safety and security of the device for shipping.

10.3 Maintenance Precautions – "Before returning the machine to operation, always check for the following". List all safety checks.

Chapter 11 - Startup Checklists

This section is most applicable to machinery. This chapter consists of a startup checklist for returning the device to service..

11.1 Startup Operator Checklist

"Prior to operating the machine, to ensure safe, smooth operation and optimum performance, closely follow this startup operator checklist."

11.1.1 Machine Startup:

11.1.2 During Operation:

Chapter 12 - Troubleshooting Guide

List the most commonly encountered problems by users (Machine will not start, cycle will not complete, etc.) and the most common steps to take to rectify the problem.

Chapter 13 – Additional Materials and Spare Parts

Provide instructions for ordering available spare parts. Provide a list of available spare parts and their part numbers.

Chapter 14 - Drawings

The drawings will be included with the Technical File. Include an index of the drawings: their titles and drawing numbers.

Documentation Overview: Technical File/Technical Construction File

If your device does not come under the governance of the Machinery Directive, you create a Technical File. If your device does come under the Machinery Directive, you create a Technical Construction File.

All test results for the device are included in the Technical File/ Technical Construction File (TF/TCF).

All other documents are included in the TF/TCF.

> *It is essential that, before drawing up the EC declaration of conformity, the manufacturer or his authorized representative established in the Community should prepare a technical construction file.*
>
> *However, it is not essential that all documentation should be permanently available in material form, but it must be possible to make it available on request. It need not include detailed plans of subassemblies used for the manufacture of machinery, unless knowledge of such plans is essential in order to ascertain conformity with the essential health and safety requirements.*

Translated into business actions, these paragraphs from the Machine Directive mean that you must create a Technical Construction File, but you do not have to include so much detail that your customer can build a replica of your machine.

The Technical File/ Technical Construction File (TF/TCF) must include a drawing (or drawings) of the complete device that shows the subassemblies and where they fit into the machine. The TF/TCF must also include enough drawing detail so that the customer can make repairs or install replacements that are likely to be necessary. The customer should be able to conduct normal maintenance duties: change worn out belts, exhausted filters, burned out bulbs, etc., by referring to the Technical Construction File for the location and access to these items.

Similarly, you must include the certifications of conformity and compliance for tests that are run on the device, but you may not need to include every page of every test, though this information must be available in a master file kept at your organization's headquarters if In Europe. Otherwise, your company must situate the complete technical and test information at a European address.

Templates/Traceability Matrices and Samples

Risk Assessment Template

#	Risk Category	Risk Description	Probability	Impact	Mitigation	Assessment	Risk Level
RA-1							
RA-2							
RA-3							
RA-4							
RA-5							

\# - Assign an in-house numbering system to track each risk. RA-1, RA-2, etc. (Risk Assessment 1 Risk Assessment 2 etc.)

Risk Category – Risks are identified and then categorized, usually by Engineering or Quality

Risk Description – Supply a brief assessment of the risk.

Probability – The levels are usually given as Frequent, Likely, Possible, Rare, Unlikely.

Impact – The levels are usually given as Catastrophic, Severe, Moderate, Minor.

Mitigation – Supply a brief description of the measures taken to bring the risk level down.

Assessment – For a product to be qualified for CE Marking, the risk must be Low, Tolerable, or Reasonable.

Risk Level - For a product to be qualified for CE Marking, the risk level should be 1, or at best a 2 with comprehensive mitigation in place.

Risk Assessment Template

Write It Yourself: CE Marking Documentation

by Jeri Massi

Declaration of Conformity Template

#	Text for the Declaration of Conformity	MACHINERY Requirement	EMC Requirement	LOW VOLTAGE Requirement
1.		Business name and full address of the manufacturer and, where appropriate, his authorised representative;	The name and address of the manufacturer and, where applicable, the name and address of his authorised representative in the Community,	Name and address of the manufacturer or his authorised representative established within the Community,
2.		Name and address of the person authorised to compile the technical file, who must be established in the Community;		
3.		Description and identification of the machinery, including generic denomination, function, model, type, serial number and commercial name;	An identification of the apparatus to which it refers, as set out in Article 9(1), *(Each apparatus shall be identified in terms of type, batch, serial number or any other information allowing for the identification of the apparatus.)*	A description of the electrical equipment

Declaration of Conformity Template

#	Text for the Declaration of Conformity	MACHINERY Requirement	EMC Requirement	LOW VOLTAGE Requirement
4.		A sentence expressly declaring that the machinery fulfils all the relevant provisions of this Directive and where appropriate, a similar sentence declaring the conformity with other Directives and/or relevant provisions with which the machinery complies. These references must be those of the texts published in the *Official Journal of the European Union*;	A reference to the Electrical Directive A dated reference to the specifications under which conformity is declared to ensure the conformity of the apparatus	Reference to the harmonised standards

Declaration of Conformity Template

#	Text for the Declaration of Conformity	MACHINERY Requirement	EMC Requirement	LOW VOLTAGE Requirement
5.		Where appropriate, the name, address and identification number of the notified body which carried out the EC type-examination referred to in Annex IX and the number of the EC type-examination certificate; Directive 2006/42/EC on Machinery foresees the involvement of a Notified Body if the product to be assessed falls in one of 23 categories listed under Annex IV and it does not conform to a European Harmonised Standard which covers all of the relevant health and safety requirement. In this case, the manufacturer has to seek assistance by a Notified Body. If it is not the case, the manufacturer himself carries out internal checks on the product.	Under the EMC conformity assessment procedures, the manufacturer is obliged to perform an EMC assessment of the apparatus. The EMC Directive does not require the intervention of a Notified Body. However, the manufacturer or his authorised representative in the Community can present technical documentation to a Notified Body which will review it and assess whether the technical documentation properly demonstrates that the requirements of the Directive have been met. If this is the case, the Notified Body will issue a statement confirming it. This statement shall be part of the technical documentation.	The manufacturer may seek the opinion of a Notified Body, but the manufacturer always has the responsibility to ensure and to declare that the products conform to the applicable legislative requirements.

Declaration of Conformity Template

#	Text for the Declaration of Conformity	MACHINERY Requirement	EMC Requirement	LOW VOLTAGE Requirement
6.		Where appropriate, the name, address and identification number of the notified body which approved the fullquality assurance system referred to in Annex X;		
7.		Where appropriate, a reference to the harmonised standards used, as referred to in Article 7(2);		
8.				Where appropriate, references to the specifications with which conformity is declared,
9.		Where appropriate, the reference to other technical standards and specifications used;		
10.				The last two digits of the year in which the CE marking was affixed
11.		The place and date of the declaration;	The date of that (this) declaration,	

Declaration of Conformity Template

#	Text for the Declaration of Conformity	MACHINERY Requirement	EMC Requirement	LOW VOLTAGE Requirement
12.		The identity and signature of the person empowered to draw up the declaration on behalf of the manufacturer or his authorised representative.	The identity and signature of the person empowered to bind the manufacturer or his authorised representative.	Identification of the signatory who has been empowered to enter into commitments on behalf of the manufacturer or his authorised representative established within the Community,

The following sample template for the DoC presents the DoC checklist template filled out with sample information.

Once you have filled out all the information in Column 2, you can pull it from the checklist, turn it into paragraphs, and paste it onto a letter head. With very few tweaks, it becomes your Declaration of Conformity.

A sample letter, pulled from the following sample checklist, follows the sample checklist.

Declaration of Conformity Requirements SAMPLE

Declaration of Conformity Checklist Sample

The EC declaration of conformity must contain the following particulars:

Text for the Declaration of Conformity	MACHINERY Requirement	EMC Requirement	LOW VOLTAGE Requirement
Business Name: X-Celeprint Address of Manufacturer: Authorized Representative:	1. business name and full address of the manufacturer and, where appropriate, his authorised representative;	the name and address of the manufacturer and, where applicable, the name and address of his authorised representative in the Community,	Name and address of the manufacturer or his authorised representative established within the Community,
Technical File Custodian: Address:	2. name and address of the person authorised to compile the technical file, who must be established in the Community;		

Declaration of Conformity Requirements SAMPLE

Text for the Declaration of Conformity	MACHINERY Requirement	EMC Requirement	LOW VOLTAGE Requirement
The X-Celeprint MTP-100 is a desktop printing device that picks and places semiconductor chips from a source substrate to a target substrate. The transfer element is a clear silicone stamp mounted on the "z" stage payload. Microscope optics are placed above the stamp for observation of stamp and sample. Point to point stage moves are coordinated by the motion controller for the transfer of chips. Model: µTP-1001 Type: µTP-100 Serial Number: Commercial Name: Micro Transfer Printer	3. description and identification of the machinery, including generic denomination, function, model, type, serial number and commercial name;	— an identification of the apparatus to which it refers, as set out in Article 9(1), ("Each apparatus shall be identified in terms of type, batch, serial number or any other information allowing for the identification of the apparatus.")	- A description of the electrical equipment,

Write It Yourself:
CE Marking Documentation

by Jeri Massi

Declaration of Conformity Requirements SAMPLE

Text for the Declaration of Conformity	MACHINERY Requirement	EMC Requirement	LOW VOLTAGE Requirement
X-Celeprint MTP-100 fulfills all the relevant provisions set forth in the following directives: DIRECTIVE 2006/42/EC OF THE EUROPEAN PARLIAMENT AND OF THE COUNCIL of 17 May 2006 on machinery, and amending Directive 95/16/EC (recast) DIRECTIVE 2004/108/EC OF THE EUROPEAN PARLIAMENT AND OF THE COUNCIL of 15 December 2004 on the approximation of the laws of the Member States relating to electromagnetic compatibility and repealing Directive 89/336/EEC DIRECTIVE 2006/95/EC OF THE EUROPEAN PARLIAMENT AND OF THE COUNCIL of 12 December 2006 on the harmonisation of the laws of Member States relating to electrical equipment designed for use within certain voltage limits	4. a sentence expressly declaring that the machinery fulfills all the relevant provisions of this Directive and where appropriate, a similar sentence declaring the conformity with other Directives and/or relevant provisions with which the machinery complies. These references must be those of the texts published in the *Official Journal of the European Union*;	a reference to this Directive, — a dated reference to the specifications under which conformity is declared to ensure the conformity of the apparatus	• Reference to the harmonised standards,

Declaration of Conformity Requirements SAMPLE

Text for the Declaration of Conformity	MACHINERY Requirement	EMC Requirement	LOW VOLTAGE Requirement
The notified body that carried out the examination referred to in Annex I of DIRECTIVE 2006/42/EC is given here: TUV Rheinland of North America, Inc. Chicago Office 2100 Golf Road, Suite 300 Rolling Meadows, IL 60008 Tel. (847) 640-5700, Fax (847) 640-5702 The number of the EC type-examination certificate: _____	5. where appropriate, the name, address and identification number of the notified body which carried out the EC type-examination referred to in Annex IX and the number of the EC type-examination certificate; Directive 2006/42/EC on Machinery foresees the involvement of a Notified Body if the product to be assessed falls in one of 23 categories listed under Annex IV and it does not conform to a European Harmonised Standard which covers all of the relevant health and safety requirement. In this case, the manufacturer has to seek assistance by a Notified Body. If it is not the case, the manufacturer himself carries out internal checks on the product..	Under the EMC conformity assessment procedures, the manufacturer is obliged to perform an EMC assessment of the apparatus. The EMC Directive does not require the intervention of a Notified Body. However, the manufacturer or his authorised representative in the Community can present technical documentation to a Notified Body which will review it and assess whether the technical documentation properly demonstrates that the requirements of the Directive have been met. If this is the case, the Notified Body will issue a statement confirming it. This statement shall be part of the technical documentation.	The manufacturer may seek the opinion of a Notified Body, but the manufacturer always has the responsibility to ensure and to declare that the products conform to the applicable legislative requirements.

Declaration of Conformity Requirements SAMPLE

Text for the Declaration of Conformity	MACHINERY Requirement	EMC Requirement	LOW VOLTAGE Requirement
The notified body that carried out the examination referred to in Annex IX of DIRECTIVE 2006/42/EC is given here: TUV Rheinland of North America, Inc. Chicago Office 2100 Golf Road, Suite 300 Rolling Meadows, IL 60008 Tel. (847) 640-5700, Fax (847) 640-5702 The number of the EC type-examination certificate:	5. where appropriate, the name, address and identification number of the notified body which carried out the EC type-examination referred to in Annex IX and the number of the EC type-examination certificate;		
N/A (X-Celeprint MTP-100 falls under the governance of Annex VIII)	6. where appropriate, the name, address and identification number of the notified body which approved the fullquality assurance system referred to in Annex X;		
N/A (X-Celeprint MTP-100 falls under the governance of Annex VIII)	7. where appropriate, a reference to the harmonised standards used, as referred to in Article 7(2);		

Declaration of Conformity Requirements SAMPLE

Text for the Declaration of Conformity	MACHINERY Requirement	EMC Requirement	LOW VOLTAGE Requirement
Detailed specifications for this device pertinent to operation, safety, and maintenance, are located in the included manual and Drawings file.			Where appropriate, references to the specifications with which conformity is declared,
The Technical File for X-Celeprint MTP-100 is included with this device in shipment and is also available for inspection at X-Celeprint business headquarters.	8. where appropriate, the reference to other technical standards and specifications used;		
The Technical File for X-Celeprint MTP-100 complies with all applicable requirements specified in DIRECTIVE 2006/42/EC, Annex VII, Part A., DIRECTIVE 2004/108/EC, and DIRECTIVE 2006/95/EC.			
The CE marking was affixed to this device in 2015.			The last two digits of the year in which the CE marking was affixed
Location of this Declaration: Date of this Declaration:	9. the place and date of the declaration;	— the date of that (this) declaration,	

Declaration of Conformity Requirements SAMPLE

Text for the Declaration of Conformity	MACHINERY Requirement	EMC Requirement	LOW VOLTAGE Requirement
Signed by: Role of Signer:	10. the identity and signature of the person empowered to draw up the declaration on behalf of the manufacturer or his authorised representative.	— the identity and signature of the person empowered to bind the manufacturer or his authorised representative.	identification of the signatory who has been empowered to enter into commitments on behalf of the manufacturer or his authorised representative established within the Community,

Declaration of Conformity Requirements SAMPLE

Sample Declaration of Conformity – Sample 2: Declaration in Letter Format

Business Name:	**ELECTRO GADGET**
Address of Manufacturer:	1133 Paradise Lane, New Meadows NC, 27605, USA
Authorized Representative:	Mike Pecci
Technical File Custodian:	Karen Custer
Address:	1133 Paradise Lane, New Meadows NC, 27605, USA

The Micro Electro Gadget 1000 is a desktop printing device that picks and places semiconductor chips from a source substrate to a target substrate. The transfer element is a clear silicone stamp mounted on the "z" stage payload. Microscope optics are placed above the stamp for observation of stamp and sample. Point to point stage moves are coordinated by the motion controller for the transfer of chips.

Model: μEG-1001
Type: μEG-100
Serial Number: EG1001100-2015
Commercial Name: Micro Electro Gadget 1000 Printer

Micro Electro Gadget 1000 fulfills all the relevant provisions set forth in the following directives:

- DIRECTIVE 2006/42/EC OF THE EUROPEAN PARLIAMENT AND OF THE COUNCIL of 17 May 2006 on machinery, and amending Directive 95/16/EC (recast)
- DIRECTIVE 2004/108/EC OF THE EUROPEAN PARLIAMENT AND OF THE COUNCIL of 15 December 2004 on the approximation of the laws of the Member States relating to electromagnetic compatibility and repealing Directive 89/336/EEC
- DIRECTIVE 2006/95/EC OF THE EUROPEAN PARLIAMENT AND OF THE COUNCIL of 12 December 2006 on the harmonisation of the laws of Member States relating to electrical equipment designed for use within certain voltage limits

The notified body that carried out the examination referred to in Annex I of DIRECTIVE 2006/42/EC is given here:

Engineering Consult of North America, Inc. Chicago Office
2100 Ruth Road, Suite 800
Arlington Heights, IL 60008
Tel. (847) 040-0000, Fax (847) 000-0002

The number of the EC type-examination certificate:	{NUMBER SUPPLIED BY NOTIFIED BODY}

page 1 of 2

Declaration of Conformity Requirements SAMPLE

Sample Declaration of Conformity – Sample 2: Declaration in Letter Format

The notified body that carried out the examination referred to in Annex IX of DIRECTIVE 2006/42/EC is given here:

Engineering Consult of North America, Inc. Chicago Office
2100 Ruth Road, Suite 800
Arlington Heights, IL 60008
Tel. (847) 040-0000, Fax (847) 000-0002

The number of the EC type-examination certificate:	NUMBER SUPPLIED BY NOTIFIED BODY

Detailed specifications for this device pertinent to operation, safety, and mainetnance, are located in the included manual and Drawings file, which make up the Technical File. The Technical File for Micro Electro Gadget 1000 is included with this device in shipment and is also available for inspection at Electro Gadget business headquarters.

The Technical File for Micro Electro Gadget 1000 complies with all applicable requirements specified in DIRECTIVE 2006/42/EC, Annex VII, Part A., DIRECTIVE 2004/108/EC, and DIRECTIVE 2006/95/EC. The CE marking was affixed to this device in 2015.

Location of this Declaration:	
Date of this Declaration:	
Signed by:	:
Printed Name of Signer:	
Role of Signer	

page 2 of 2

User Manual Requirements Template

User Manual Requirements Checklist/Traceability Matrix

The manual is the largest written document in the CE Marking documentation set. A sample manual outline has been provided previously in this handbook. One other strategy to make the CE Marking project easier and more trackable is to use the checklist as your traceability matrix. Instead of merely checking off each item as it is written into the user's manual, use the second column in the checklist to enter the section number of the manual where you have entered the required information from the Directive(s) entries in columns three, four, and/or five.

The left (first) column of the table below supplies you with a Traceability Matrix numbering system (TM-Mnn). Enter the section number from the manual in the next column, and you will have created traceability from the manual you are creating to the Directives so that you can track the development of your project and make sure that you cover all requirements from the Directives.

Note: If your device does not fall under all three directives, you may edit the checklist or simply cross out any non-pertinent sections. You may scan the following pages and edit them to tweak them for your device. But you may not sell copies or hand them out in any paid venue, per copyright law.

> *Some items required for the User Manual may be presented in the Technical File or even the Risk Assessment. If so, note that in the check box (TF or RA). Refer to the sample checklist that follows this document.

User Manual Requirements Template

User Manual Requirements Template

Number TM-NNN	Section Number (Manual)*	CELEX-MACHINERY- 32006L0042-EN-TXT-see Annex 1-1.7.4.1 MACHINERY	CELEX-VOLTAGE 32006L0095-EN-TXT LOW VOLTAGE DEVICES	CELEX-ELECTRICAL- 32004L0108-EN-TXT emc/emf
TM-M001		1.7.4.1-a The instructions must be drafted in one or more official Community languages.		
TM-M002		1.7.4.1-a The words 'Original instructions' must appear on the language version(s) verified by the manufacturer or his authorised representative.		
TM-M003		1.7.4.1-b Where no 'Original instructions' exist in the official language(s) of the country where the machinery is to be used, a translation into that/those language(s) must be provided by the manufacturer or his authorised representative or by the person bringing the machinery into the language area in question. The translations must bear the words 'Translation of the original instructions'.		
TM-M004			A4-3 (General) Technical documentation must enable the conformity of the electrical equipment to the requirements of this Directive to be assessed.	Annex 4.1 The technical documentation must enable the conformity of the apparatus with the essential requirements to be assessed. It must cover the design and manufacture of the apparatus, in particular:

User Manual Requirements Template

Number TM-NNN	Section Number (Manual)*	CELEX-MACHINERY-32006L0042-EN-TXT-see Annex1-1.7.4.1 MACHINERY	CELEX-VOLTAGE 32006L0095-EN-TXT LOW VOLTAGE DEVICES	CELEX-ELECTRICAL-32004L0108-EN-TXT emc/emf
TM-M005		1.7.4.2-a The business name and full address of the manufacturer and of his authorised representative;		
TM-M006		1.7.4.2-b The designation of the machinery as marked on the machinery itself, except for the serial number (see section 1.7.3 of Machinery Directive);		
TM-M007		1.7.4.2-c The EC declaration of conformity, or a document setting out the contents of the ec declaration of conformity, showing the particulars of the machinery, not necessarily including the serial number and the Signature;		
TM-M008		1.7.4.2-d A general description of the machinery;	A4-3 A general description of the electrical equipment,	Annex 4 A general description of the apparatus;
TM-M009				Annex 4 Documentation: Good Engineering Practices have been followed

User Manual Requirements Template

Number TM-NNN	Section Number (Manual)*	CELEX-MACHINERY-32006L0042-EN-TXT-see Annex1-1.7.4.1 MACHINERY	CELEX-VOLTAGE 32006L0095-EN-TXT LOW VOLTAGE DEVICES	CELEX-ELECTRICAL-32004L0108-EN-TXT emc/emf
TM-M010		1.7.4.2-e The drawings, diagrams, descriptions and explanations necessary for the use, maintenance and repair of the machinery and for checking its correct functioning;	A4-3 Conceptual design and manufacturing drawings and schemes of components, sub-assemblies, circuits, etc., Descriptions and explanations necessary for the understanding of said drawings and schemes and the operation of the electrical equipment,	
TM-M011		1.7.4.2-f A description of the workstation likely to be occupied by operators;		
TM-M012		1.7.4.2-g A description of the intended use of the machinery;		
TM-M013		1.7.4.2-h Warnings concerning ways in which the machinery must not be used that experience has shown might occur;		
TM-M014		1.7.4.2-i Assembly, installation and connection instructions, including drawings, diagrams and the means of attachment		
TM-M015		1.7.4.2-i The designation of the chassis or installation on which the machinery is to be mounted;		

User Manual Requirements Template

Number TM-NNN	Section Number (Manual)*	CELEX-MACHINERY-32006L0042-EN-TXT-see Annex1-1.7.4.1 MACHINERY	CELEX-VOLTAGE 32006L0095-EN-TXT LOW VOLTAGE DEVICES	CELEX-ELECTRICAL-32004L0108-EN-TXT emc/emf
TM-M016		1.7.4.2-j Instructions relating to installation and assembly for reducing noise or vibration;		
TM-M017		1.7.4.2-k Instructions for the putting into service and use of the machinery and,		
TM-M018		1.7.4.2-k If necessary, instructions for the Training of operators;		
TM-M019		1.7.4.2-l Information about the residual risks that remain despite the inherent safe design measures, safeguarding		
TM-M020		1.7.4.2-l And complementary protective measures adopted;		
TM-M021		1.7.4.2-m Instructions on the protective measures to be taken by the user, including, where appropriate, the		
TM-M022		1.7.4.2-m Personal protective equipment to be provided;		
TM-M023		1.7.4.2-n The essential characteristics of tools which may be fitted to the machinery;		

by Jeri Massi

Write It Yourself:
CE Marking Documentation

User Manual Requirements Template

Number TM-NNN	Section Number (Manual)*	CELEX-MACHINERY-32006L0042-EN-TXT-see Annex1-1.7.4.1 MACHINERY	CELEX-VOLTAGE 32006L0095-EN-TXT LOW VOLTAGE DEVICES	CELEX-ELECTRICAL-32004L0108-EN-TXT emc/emf
TM-M024		1.7.4.2-o The conditions in which the machinery meets the requirement of stability during the following: -Use, -Transportation, -Assembly, -Dismantling when out of service, testing or foreseeable breakdowns;		
TM-M025		1.7.4.2-p Instructions with a view to ensuring that transport, handling and storage operations can be made safely,		
TM-M026		1.7.4.2-p Giving the mass of the machinery and of its various parts where these are regularly to be transported separately;		
TM-M027		1.7.4.2-q The operating method to be followed in the event of accident or breakdown;		
TM-M028		1.7.4.2-q If a blockage is likely to Occur, the operating method to be followed so as to enable the equipment to be safely unblocked;		

User Manual Requirements Template

Number TM-NNN	Section Number (Manual)*	X-MACHINERY-32006L0042-EN-TXT-see Annex1-1.7.4.1 MACHINERY	CELEX-VOLTAGE 32006L0095EN-TXT W VOLTAGE DEVICES	EX-ELECTRICAL-32004L0108-EN-TXT emc/emf
TM-M029		1.7.4.2-r The description of the adjustment and maintenance operations that should be carried out by the user and		
TM-M030		1.7.4.2-r The preventive maintenance measures that should be observed;		
TM-M031		1.7.4.2-s Instructions designed to enable adjustment and maintenance to be carried out safely, including the		
TM-M032		1.7.4.2-s Protective measures that should be taken during these operations;		
TM-M033		1.7.4.2-t The specifications of the spare parts to be used, when these affect the health and safety of operators;		

Write It Yourself:
CE Marking Documentation

54

by Jeri Massi

Number TM-NNN	Section Number (Manual)*			
TM-M034		X-MACHINERY-32006L0042-EN-TXT-see Annex1-1.7.4.1 MACHINERY	CELEX-VOLTAGE 32006L0095EN-TXT LOW VOLTAGE DEVICES	EX-ELECTRICAL-32004L0108-EN-TXT emc/emf
		1.7.4.2-u The following information on airborne noise emissions: 1) the a-weighted emission sound pressure level at workstations, where this exceeds 70 db; where 2) This level does not exceed 70 db, this fact must be indicated, 3)the peak c-weighted instantaneous sound pressure value at workstations, where this exceeds 63 pa (130 db in relation to 20 μpa), 4) the a-weighted sound power level emitted by the machinery, where the a-weighted emission sound, 5) Pressure level at workstations exceeds 80 db.		

User Manual Requirements Template

Number TM-NNN	Section Number (Manual)*	CELEX-MACHINERY-32006L0042-EN-TXT-see Annex1-1.7.4.1 MACHINERY	CELEX-VOLTAGE 32006L0095-EN-TXT LOW VOLTAGE DEVICES	CELEX-ELECTRICAL-32004L0108-EN-TXT emc/emf
TM-M035			A4-3 A list of the standards applied in full or in part, and descriptions of the solutions adopted to satisfy the safety aspects of this Directive where standards have not been applied,	
TM-M036			A4-3 Results of design calculations made, examinations carried out, etc.,	
TM-M037			A4-3 Test reports. (See *Technical File.*)	Annex 4 a description of the electromagnetic compatibility assessment set out in Annex II, point 1, results of design calculations made, examinations carried out, test reports, etc.;
TM-M038				Annex 1.a Documentation: (a) the electromagnetic disturbance generated does not exceed the level above which radio and telecommunications equipment or other equipment cannot operate as intended;

User Manual Requirements Template

Number TM-NNN	Section Number (Manual)*	CELEX-MACHINERY-32006L0042-EN-TXT-see Annex1-1.7.4.1 MACHINERY	CELEX-VOLTAGE 32006L0095-EN-TXT LOW VOLTAGE DEVICES	CELEX-ELECTRICAL-32004L0108-EN-TXT emc/emf
TM-M039				Annex 1.b Documentation: (b) it has a level of immunity to the electromagnetic disturbance to be expected in its intended use which allows it to operate without unacceptable degradation of its intended use.
TM-M040				Annex 4 Evidence of compliance with the harmonised standards, if any, Annex 4 applied in full or in part
TM-M041				Annex 4 Where the manufacturer has not applied harmonised standards, or has applied them only in part, a description and explanation of the steps taken to meet the essential requirements of the Directive, including a description of the electromagnetic compatibility assessment set out in Annex II, point 1, results of design calculations made,
TM-M042		1.7.4.1-(c) The contents of the instructions must cover not only the intended use of the machinery but also take into account any reasonably foreseeable misuse thereof.		

User Manual Requirements Template

Number TM-NNN	Section Number (Manual)*	CELEX-MACHINERY-32006L0042-EN-TXT-see Annex1-1.7.4.1 MACHINERY	CELEX-VOLTAGE 32006L0095-EN-TXT LOW VOLTAGE DEVICES	CELEX-ELECTRICAL-32004L0108-EN-TXT emc/emf
TM-M043		1.7.4.1-(d) In the case of machinery intended for use by non-professional operators, the wording and layout of the instructions for use must take into account the level of general education and acumen that can reasonably be expected from such operators.		

*Some items may be presented in the Technical File or even the Risk Assessment. If so, note that in the check box (TF or RA). Refer to the sample checklist that follows this document.

Notes

User Manual Requirements Sample

The following pages provide a sample of the User Manual Requirements Checklist, as it would be used during a project. The writer must track *where* (as in, in which document), each requirement has been met. That information is entered, in abbreviated format, in Column 2.

User Manual Requirements Sample

User Manual Requirements Checklist/Traceability Matrix Sample

User Manual Requirements Sample

Number TM-NNN	Section Number (Manual)	CELEX-MACHINERY-32006L0042-EN-TXT-see Annex1-1.7.4.1 MACHINERY	CELEX-VOLTAGE 32006L0095-EN-TXT LOW VOLTAGE DEVICES	CELEX-ELECTRICAL-32004L0108-EN-TXT emc/emf
TM-M001	MAN	1.7.4.1-a The instructions must be drafted in one or more official Community languages.		
TM-M002	inside cover	1.7.4.1-a The words 'Original instructions' must appear on the language version(s) verified by the manufacturer or his authorised representative.		
TM-M003	N/A	1.7.4.1-b Where no 'Original instructions' exist in the official language(s) of the country where the machinery is to be used, a translation into that/those language(s) must be provided by the manufacturer or his authorised representative or by the person bringing the machinery into the language area in question. The translations must bear the words 'Translation of the original instructions'.		
TM-M004	See this manual, Tech File, DoC		A4-3 (*General*) Technical documentation must enable the conformity of the electrical equipment to the requirements of this Directive to be assessed.	Annex 4.1 The technical documentation must enable the conformity of the apparatus with the essential requirements to be assessed. It must cover the design and manufacture of the apparatus, in particular:

Write It Yourself:
CE Marking Documentation

62

by Jeri Massi

User Manual Requirements Sample

TM-M005	page 2	1.7.4.2-a The business name and full address of the manufacturer and of his authorised representative;	
TM-M006	Sec 1.3	1.7.4.2-b The designation of the machinery as marked on the machinery itself, except for the serial number (see section 1.7.3 of *Machinery Directive*);	
TM-M007	See DoC	1.7.4.2-c The EC declaration of conformity, or a document setting out the contents of the ec declaration of conformity, showing the particulars of the machinery, not necessarily including the serial number and the Signature;	
TM-M008	Sec 1.1	1.7.4.2-d A general description of the machinery;	A4-3 A general description of the electrical equipment,	Annex 4 A general description of the apparatus;
TM-M009	See Tech File		
TM-M010	See Tech File, also chps 4-13 OPEN	1.7.4.2-e The drawings, diagrams, descriptions and explanations necessary for the use, maintenance and repair of the machinery and for checking its correct functioning;	A4-3 Conceptual design and manufacturing drawings and schemes of components, sub-assemblies, circuits, etc., Descriptions and explanations necessary for the understanding of said drawings and schemes and the operation of the electrical equipment.	Annex 4 Documentation: Good Engineering Practices have been followed

User Manual Requirements Sample

TM-M011	Sec 3.2	1.7.4.2-f A description of the workstation likely to be occupied by operators;
TM-M012	Secs 1.1, 1.2	1.7.4.2-g A description of the intended use of the machinery;
TM-M013	Chp 2	1.7.4.2-h Warnings concerning ways in which the machinery must not be used that experience has shown might occur;
TM-M014	Chp 3 OPEN	1.7.4.2-i Assembly, installation and connection instructions, including drawings, diagrams and the means of attachment
TM-M015	1.1.3 and 3.2	1.7.4.2-i The designation of the chassis or installation on which the machinery is to be mounted;
TM-M016	3.2	1.7.4.2-j Instructions relating to installation and assembly for reducing noise or vibration;
TM-M017	Chp. 3	1.7.4.2-k Instructions for the putting into service and use of the machinery and,
TM-M018	2.4	1.7.4.2-k If necessary, instructions for the Training of operators;
TM-M019	Chp 2	1.7.4.2-l Information about the residual risks that remain despite the inherent safe design measures, safeguarding
TM-M020	Chp 2	1.7.4.2-l And complementary protective measures adopted;
TM-M021	Secs 2.4; 3.5	1.7.4.2-m Instructions on the protective measures to be taken by the user, including, where appropriate, the

Write It Yourself:
CE Marking Documentation

64

by Jeri Massi

User Manual Requirements Sample

TM-M022	N/A	1.7.4.2-m Personal protective equipment to be provided;
TM-M023	Sec 2.4	1.7.4.2-n The essential characteristics of tools which may be fitted to the machinery;
TM-M024	Sec 3.5; 9.2	1.7.4.2-o The conditions in which the machinery meets the requirement of stability during the following: -Use, -Transportation, -Assembly, -Dismantling when out of service, testing or foreseeable breakdowns;
TM-M025	Sec 3.4, 3.5, and 9.2	1.7.4.2-p Instructions with a view to ensuring that transport, handling and storage operations can be made safely,
TM-M026	Secs 1.3; 3.5	1.7.4.2-p Giving the mass of the machinery and of its various parts where these are regularly to be transported separately;
TM-M027	2.4	1.7.4.2-q If a blockage is likely to Occur, the operating method to be followed in the event of accident or breakdown;
TM-M028	2.4	1.7.4.2-q The operating method to be followed so as to enable the equipment to be safely unblocked;
TM-M029	Chps 8, 9, 10 OPEN	1.7.4.2-r The description of the adjustment and maintenance operations that should be carried out by the user and
TM-M030	Chps 8, 9, 10	1.7.4.2-r The preventive maintenance measures that should be observed;

User Manual Requirements Sample

TM-M031	Chps 9	1.7.4.2-s Instructions designed to enable adjustment and maintenance to be carried out safely, including the
TM-M032	Chp 2	1.7.4.2-s Protective measures that should be taken during these operations;
TM-M033	Chp. 12	1.7.4.2-t The specifications of the spare parts to be used, when these affect the health and safety of operators;
TM-M034	Appendices OPEN	1.7.4.2-u The following information on airborne noise emissions: 1) the a-weighted emission sound pressure level at workstations, where this exceeds 70 db; where 2) This level does not exceed 70 db, this fact must be indicated, 3)the peak c-weighted instantaneous sound pressure value at workstations, where this exceeds 63 pa (130 db in relation to 20 µpa), 4) the a-weighted sound power level emitted by the machinery, where the a-weighted emission sound, 5) Pressure level at workstations exceeds 80 db.

Write It Yourself:
CE Marking Documentation

User Manual Requirements Sample

TM-M035	Chp 2		A4-3 A list of the standards applied in full or in part, and descriptions of the solutions adopted to satisfy the safety aspects of this Directive where standards have not been applied,
TM-M036	Appendices Tech File OPEN		A4-3 Results of design calculations made, examinations carried out, etc.,
TM-M037	Appendices Tech File OPEN		A4-3 Test reports. (See Technical File.)
TM-M038	Appendices Tech File OPEN		Annex 4 a description of the electromagnetic compatibility assessment set out in Annex II, point 1, results of design calculations made, examinations carried out, test reports, etc.;
TM-M039	Appendices Tech File OPEN		Annex 1.a Documentation: (a) the electromagnetic disturbance generated does not exceed the level above which radio and telecommunications equipment or other equipment cannot operate as intended; Annex 1.b Documentation: (b) it has a level of immunity to the electromagnetic disturbance to be expected in its intended use which allows it to operate without unacceptable degradation of its intended use.

User Manual Requirements Sample

TM-M040	DoC OPEN		Annex 4 Evidence of compliance with the harmonised standards, if any, Annex 4 applied in full or in part
TM-M041	Appendices Tech File OPEN		Annex 4 Where the manufacturer has not applied harmonised standards, or has applied them only in part, a description and explanation of the steps taken to meet the essential requirements of the Directive, including a description of the electromagnetic compatibility assessment set out in Annex II, point 1, results of design calculations made, COULD BE REDUNDANT
TM-M042	Chp 2	1.7.4.1-(c) The contents of the instructions must cover not only the intended use of the machinery but also take into account any reasonably foreseeable misuse thereof.	
TM-M043	User Manual	1.7.4.1-(d) In the case of machinery intended for use by non-professional operators, the wording and layout of the instructions for use must take into account the level of general education and acumen that can reasonably be expected from such operators.	

Write It Yourself:
CE Marking Documentation

68

by Jeri Massi

Notes

Technical File/Technical Construction File (TF/TCF) Template

The Technical File/Technical Construction File (TF/TCF)

The Machinery Directive specifies that "the technical file must demonstrate that the machinery complies with the requirements of this Directive. It must cover the design, manufacture and operation of the machinery to the extent necessary for this assessment." Essentially, this means that all the documentation becomes part of the Technical File/Technical Construction File (TF/TCF).

Realistically, you will work on each document separately and then put them in order at the end of the project as components of the TF/TCF. However, the component that is unique to the TF/TCF is the drawings package. As is stated earlier in this handbook, you don't have to (and definitely should not) include all the drawings that make up you product. But you do need to include those drawings and schematics and Bills of Materials that the customer will need for operation, for routine maintenance, and for non-proprietary repairs (changing light bulbs, replacing filters, replacing belts, etc.)

For ease of use, the checklist for just the Machinery Technical Construction File will be presented first. It is far more extensive than the requirements for the Low Voltage or EMF devices. A combined checklist for the Low Voltage and EMF devices will be presented after the Machinery checklist.

Technical File/Technical Construction File (TF/TCF) Template

Machinery Technical Construction File Traceability Matrix

Number	Where to find or N/A	The technical file shall include the following:
TM-T01		A construction file including: — a general description of the machinery,
TM-T02		A construction file including: — the overall drawing of the machinery and drawings of the control circuits, as well as the pertinent descriptions and explanations necessary for understanding the operation of the machinery,
TM-T03		A construction file including: — full detailed drawings,
TM-T04		A construction file including: accompanied by any calculation notes, required to check the conformity of the machinery with the essential health and safety requirements,
TM-T05		A construction file including: test results, required to check the conformity of the machinery with the essential health and safety requirements,
TM-T06		A construction file including: certificates, etc., required to check the conformity of the machinery with the essential health and safety requirements,
TM-T07		Risk Assessment - a list of the essential health and safety requirements which apply to the machinery,
TM-T08		Risk Assessment - the description of the protective measures implemented to eliminate identified hazards or to reduce Risks and, when appropriate, the indication of the residual risks associated with the machinery,
TM-T09		— The standards and other technical specifications used, indicating the essential health and safety requirements covered by these standards
TM-T10		— Any technical report giving the results of the tests carried out either by the manufacturer or by a body chosen by the manufacturer or his authorized representative,
TM-T11		— A copy of the instructions for the machinery,
TM-T12		— Where appropriate, the declaration of incorporation for included partly completed machinery and the relevant assembly instructions for such machinery,
TM-T13		— where appropriate, copies of the EC declaration of conformity of machinery or other products incorporated into the machinery,
TM-T14		— a copy of the EC declaration of conformity;

Technical File/Technical Construction File (TF/TCF) Template

Low Voltage/EMF Technical File Traceability Matrix

Number TM-NNN	Location or Tech File (TF)	CELEX-VOLTAGE 32006L0095-EN-TXT LOW VOLTAGE DEVICES	CELEX-ELECTRICAL- 32004L0108-EN-TXT emc/emf
TM-TT001		A4-3 *(General)* Technical documentation must enable the conformity of the electrical equipment to the requirements of this Directive to be assessed.	Annex 4.1 *(General)* The technical documentation must enable the conformity of the apparatus with the essential requirements to be assessed. It must cover the design and manufacture of the apparatus, in particular:
TM-TT002			Annex 4 *(General)* Documentation: Good Engineering Practices have been followed
TM-TT003		A4-3 Conceptual design and manufacturing drawings and schemes of components, sub-assemblies, circuits, etc., Descriptions and explanations necessary for the understanding of said drawings and schemes and the operation of the electrical equipment,	
TM-TT004		A4-3 Results of design calculations made, examinations carried out, etc.,	
TM-TT005		A4-3 Test reports.	Annex 4 a description of the electromagnetic compatibility assessment set out in Annex II, point 1, results of design calculations made, examinations carried out, test reports, etc.;

Write It Yourself: CE Marking Documentation — by Jeri Massi

Technical File/Technical Construction File (TF/TCF) Template

Number TM-NNN	Location or Tech File (TF)	CELEX-VOLTAGE 32006L0095-EN-TXT LOW VOLTAGE DEVICES	CELEX-ELECTRICAL- 32004L0108-EN-TXT emc/emf
TM-TT006			Annex 1.a Documentation: (a) the electromagnetic disturbance generated does not exceed the level above which radio and telecommunications equipment or other equipment cannot operate as intended;
TM-TT007			Annex 1.b Documentation: (b) it has a level of immunity to the electromagnetic disturbance to be expected in its intended use which allows it to operate without unacceptable degradation of its intended use.
TM-TT008			Annex 4 Where the manufacturer has not applied harmonized standards, or has applied them only in part, a description and explanation of the steps taken to meet the essential requirements of the Directive, including a description of the electromagnetic compatibility assessment set out in Annex II, point 1, results of design calculations made,

Electrical Component File

The Electrical Component File

The Electrical Component File (or list) is not explicitly required by the directives, but it supplies up-front information on the quality of your company's machine or device. It is a list, draw from the BOMs of the assemblies that make up the device, of the electrical components of the system, with a record of the quality standards that they meet.

Sadly, this means that the technical writer must go through the drawings and the BOMs to look up each electrical component, write it down, and then check the documentation that came with it from the vendor to find the quality markings, such as UL, CSA, IEC, CE, ROHS, etc., and document them. But having a component genealogy that shows that quality standards have been applied down to the component level is one way to prove the quality, reliability, and safety of the device. The following sample is brief, but an ECF file can go on for pages for larger devices.

The format of the ECF file is so simple that the easiest way to use it from this document is to either scan it and convert it to a word processing program and use it electronically, or enter it manually into a word processor or spreadsheet program and work from that.

Electrical Component File

SAMPLE

Type of Product: Electric Whirly Gadget

Model Number: EWG-1001

List of Critical Components:

Object/part no.	Manufacturer/ Trademark	Type/Model	Technical Data	Standard	Mark(s) of Conformity
Load Switch	Allen Bradley	194E-E16-1753-4N	Voltage Rating: 600VAC, Current Rating: 16A Max	UL 508, CSA C22.2, No. 14, IEC 60947-3, CE	UL, CSA, IEC, CE
Dual Output Lighting Controller	Advanced Illumination	SL4301A-WHIC2	24VDC	CE, ROHS	CE, ROHS
Ensemble MP Controller And PWM Digital Drive	Aerotech	MP10	FIREWIRE, 10A, 5A, +/-40 VDC, PWM, 1-AI, 8-DO/8DI, 1-AO/1-AI	NRTL Safety Certification, Ce Approved	CE, NRTL
Safety Relay Module	Altech	8956.2C	4 DPDT Relays 24VDC Coil 250VAC 8A Contacts DIN RAIL MT	TÜV, UL, OSHA FR19, EMD 89.392 EEC	TÜV, UL, OSHA, EMD 89.392, EEC

Electrical Component File

Template

Type of Product: Electric Whirly Gadget

Model Number: EWG-1001

List of Critical Components:

Object/part no.	Manufacturer/ Trademark	Type/Model	Technical Data	Standard	Mark(s) of Conformity

Electrical Component File

Terms

CEN stands for European Committee for Standardization.
CENELEC stands for European Committee for Electro technical Standardization.

Printed in Great Britain
by Amazon